THE CHANGING FACE OF THE CZECH REPUBLIC

Text by JACOB RIHOSEK
Photographs by JENNY MATTHEWS

Chicago, Illinois

Published by Raintree, a division of Reed Elsevier, Inc.
Chicago, Illinois
Customer Service 888-363-4266
Visit our website at www.raintreelibrary.com

For information address the publisher:
Raintree
100 N. LaSalle
Suite 1200
Chicago, IL 60602

Library of Congress Cataloging-in-Publication Data:

Řihošek, Jacob.
Czech Republic / Jacob Řihošek.
p. cm. -- (The changing face of--)
Summary: Presents the natural environment and resources, people and culture, and business and economy of the Czech Republic, focusing on development and change in recent years.
Includes bibliographical references and index.
ISBN 0-7398-6828-4
1. Czech Republic--Juvenile literature. [1. Czech Republic.] I. Title: Changing face of Czech Republic. II. Title. III. Series.
DB2065.R54 2004
943.7105--dc21

2003009747

Printed in China by WKT

08 07 06 05 04
10 9 8 7 6 5 4 3 2 1

Acknowledgments
The publishers would like to thank the following for their contributions to this book: Rob Bowden—statistics research; Peter Bull—map illustration; Nick Hawken—statistics panel illustrations. Thanks also to Tomas Řihošek for translations and Jenny Matthews for the interviews. All photographs are by Jenny Matthews except: p. 9 (top) Joe Klamar/Reuters/Popperfoto.com; p. 10 (top) Corbis/Jonathan Blair; p. 23 (top) Corbis/Liba Taylor; p.33 (top) Sue Ogrocki/Reuters/Popperfoto.com.

Contents

The Capital

For more than a thousand years, Prague has been the biggest city and the center of government of the country now known as the Czech Republic. In ancient times Prague was home to the Czech royalty. Today the president lives in Hradcany Castle, which overlooks the Vltava, the city's main river. The city's location, in the center of the country, has played an important role in its trade. Prague's markets were renowned as the biggest and the richest in the country. Wenceslas Square, named after the first Christian ruler of the Czechs, was once home to a horse fair; today it is one of the city's biggest tourist attractions.

▲ *Hradcany Castle in Prague now serves as a presidential palace. A special flag is raised when the president is present.*

For much of the 20th century, the ruling Communist Party made access to Prague difficult for some foreign visitors. However, since major political changes occurred after the fall of the communist government in 1989 (see page 6), tourism has become one of the Czech Republic's main industries. Visitors from all over the world come to see Prague's stunning architecture and to view the superb collection of art in its many galleries.

Most of the changes seen in Prague since 1989 have begun to spread to the rest of the country; foreign businesses are investing in the Czech Republic and the country is undergoing great modernization and change.

◀ *A supermarket building in central Prague. After 1989 many of the capital's streets were modernized.*

▲ *This map shows the main geographical features of the Czech Republic as well as most of the places mentioned in this book.*

THE CZECH REPUBLIC: KEY FACTS

Area: 30,450 square miles (78,866 square kilometers)
Population: 10.3 million
Population density: 50 people per square mile (130 people per square kilometer)
Capital city: Prague
Other main cities: Brno, Plzen, Ostrava (each with less than one million residents)
Highest mountain: Sněžka (5,256 feet [1,602 meters])
Longest river: Vltava (267 miles [430 kilometers])
Main language: Czech
Major religions: Atheist 39.8 percent, Roman Catholic 39.2 percent, Protestant 4.6 percent, Orthodox 3 percent, other 13.4 percent
Currency: Czech crown (1 crown = 100 halers)

2 Past Times

The Czech Republic was once part of Czechoslovakia, a federation (combination) of the Czechs and the Slovaks. The two countries joined in 1918. In 1948, Czechoslovakia became part of the Eastern Bloc, a group of countries dominated by the Soviet Union. Little contact was allowed with Western Europe and all businesses were controlled by the communist government, which remained in power from 1948 until 1989.

▲ *Statues of workers from the post–war era can be seen in the Museum of Communism in Prague.*

During this period most Czechs and Slovaks were restricted by the government in many areas of their lives. For example, it wasn't easy to get a good job without being a member of the Communist Party. People could not freely express different political views without the danger of being prosecuted by the government. The year 1968 was the most difficult period of the postwar history of the Czech Republic. The invading Soviet armies repressed a strong movement by the people, against the authority of the Soviet Union. Although there was no armed conflict, many activists of the movement were imprisoned, and people became afraid to raise their voices. In the late 1980s the power of the Soviet Union was beginning to decrease and many Czech and Slovak people decided that it was time for a change. In 1989, after a peaceful street protest in Prague, the Communist Party gave way to a more democratic system. This marked an important change in Czechoslovakian politics and economy.

▼ *The famous astronomical clock of Prague. This major tourist attraction was built in 1410 C.E. and is still functioning today.*

New opportunities opened up for trading with countries in the West. The effects on Czech society were far reaching. For example, differences between social classes became more apparent and have continued to increase. The latest major change in Czech history took place in 1992, when, after a mutual agreement, the federation of Czechoslovakia was split into the Czech Republic and the Slovak Republic, as they are today.

▶ *This British supermarket branch in central Prague is an example of how foreign businesses expanded in the Czech Republic after 1989.*

IN THEIR OWN WORDS

My name is Jan Řihošek. I am optimistic for our children. They will be a happier generation without the experience of 1948 when the communist government took over or 1968 when the Soviets occupied us. I was so surprised when the revolution happened in 1989. I never thought it would be possible to change the socialist regime.

Now there are great changes happening in our country. One positive change is that children now have good environmental education. I run a sports shop that sells skateboards, snowboards, windsurfing boards, and sailing equipment. Western sports have become very popular with the younger generation. However, even in these changing times, it is hard to survive economically because I have to compete with other shops and I rely on young people having enough money to spend.

3 Landscape and Climate

The Czech Republic is a small, landlocked country in the heart of the European continent. It is located on some of the oldest and most significant land routes in Europe. The country doesn't have any coastline, but it shares its borders with four other countries: Poland in the north, Slovakia in the east, Austria in the south, and Germany in the west.

◀ *The lack of coastline means that all the areas of water in the Czech Republic are freshwater. Most of the slow-moving sections of the rivers freeze over when temperatures drop in winter.*

The seasons

From the north of the Czech Republic to the south is only a few hundred miles, so the weather conditions and temperatures are very similar all over the country. The weather in the Czech Republic is affected by two very different climates. The weather systems coming from the west bring humid weather from the Atlantic Ocean, and the easterly winds from Asia are drier and very cold in winter. There are four clearly defined seasons. The normal summer temperature is around 79 °F (26 °C). In winter the temperature varies between 12 °F and 32 °F (–11 °C and 0 °C), although the temperatures are even lower in the mountains.

▼ *Spring in the Czech Republic brings a dramatic change to the landscape. The bare trees and yellow grass are replaced by abundant greenery.*

Flooding

In recent years there has been more rain in the summer months than ever before. Scientists believe that this is happening because the world climate is becoming warmer. As a result, the Czech Republic has recently experienced the most destructive floods in its history. Floods that have occurred in the past few years have caused billions of dollars worth of damage. During the floods in 2002, which also affected Germany and Austria, the water level rose so high in Prague that it flooded the subway system. Many people were also left homeless after their houses collapsed from the flood.

▶ *The floods in Prague in 2002 caused the Vltava River to rise dramatically, which badly damaged the city.*

IN THEIR OWN WORDS

I'm Tomas Řihošek. I was born in Olomouc, but I have been living in Prague for the last two years. The climate in the Czech Republic is pretty balanced, with warm summers and cold, snowy winters. People say it is getting warmer, but it is hard to tell. I have seen two floods, in 1997 and in 2002. In Olomouc in 1997, it rained for three months and flooded my dad's shop and nearly ruined his business. In 2002 people took more notice because it affected Prague and transportation was badly disrupted; the commuter train line closed down for six months. Around the Vltava River in Prague, drainage tunnels are being built. Some of the flooding has occurred because forests have been cut down. The soil has eroded away and rivers have been diverted from their natural courses. This is so that farmers have dry fields all year round and housing can be built on the floodplain.

Forests

Forests cover 34 percent of the Czech Republic. There are areas, especially in the north, where trees and soil have suffered damage from acid rain. The government has taken big steps to try to improve the quality of the air, including the introduction of stricter laws to reduce factory emissions. The levels of sulphur dioxide, the main chemical responsible for acid rain, are now ten times less than they were in 1987.

▲ *Decades of acid rain have destroyed these Norway spruce trees. It could take more than 200 years for the soil to recover.*

Rivers and lakes

The Czech Republic has many lakes and rivers. Only a few lakes are natural. These lakes, called glacial lakes, are in the mountains. Glacial lakes were created during the last Ice Age when glaciers pushed through the valleys and melted into lakes. Most lakes are resevoirs for drinking water and were created by dams. The dams are also used to generate hydroelectricity. The biggest dam, Lipno, is on the Vltava River in the south of the country.

The Labe River is not as long as the Vltava, but it is wider and deeper. It provides a shipping connection between the port of Ústí nad Labem in the north and the North Sea.

▶ *Summertime is a welcome opportunity for Prague's visitors to explore the Vltava River in rented boats.*

Mountains

There are many different mountain ranges, mostly along the border regions. The highest mountain is Snežka at 5,256 ft (1,602 m), in the Krkonoše Mountains in the north. In the southwest, the Sumava range is where the Vltava River has its source. There are many national reserves in the mountains with scenic landscapes, castle ruins, and a lot of beautiful woodland and protected wildlife species. Most of the Czech Republic's forests are in the mountains, and they consist mainly of coniferous trees.

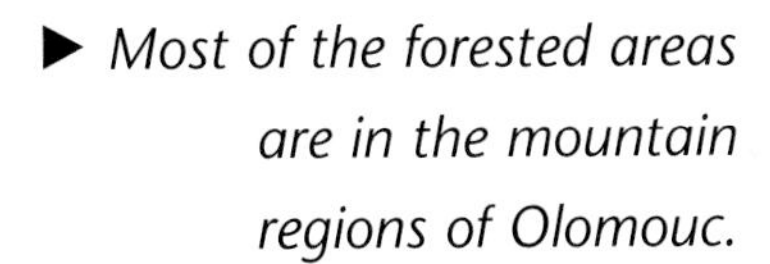

▶ *Most of the forested areas are in the mountain regions of Olomouc.*

IN THEIR OWN WORDS

I'm Eva Martinková, and I belong to the Czech Union of Environmentalists, which is a nongovernmental organization that arranges activities such as cleaning up rivers and planting new forests. If a law is broken, groups like ours are allowed to intervene and to demand settlement. For example, yesterday I came across a man who had plowed a field where there is a type of protected iris. I told him he had to return the meadow to its original state or he would be taken to court.

The main change to the Czech environment is that controls on air pollution caused by the coal-power plants have been introduced. Although there are more cars causing pollution, they are mainly in the cities and on the highways, whereas the coal plants were located across the whole country.

4 Natural Resources

Energy sources

In the Czech Republic most energy is currently produced from coal. The coal is burned in huge quantities to provide the country with more than 70 percent of its electricity. Nuclear power production is not as developed and provides only 20 percent of energy. The Czech Republic produces only small quantities of oil and mainly relies on imported oil from Russia. A pipeline (the Ingolstadt–Kralupy pipeline) has been built to carry oil from the Adriatic Sea to the Czech Republic. Other energy sources include hydroelectricity, wind, and solar power. These alternative energy sources are the most ecological, but they currently produce only a fraction of the energy needed. This is because they only began to be used after the fall of the communist government and they are still being developed.

▲ *Most of the electricity in the Czech Republic is produced by coal-powered plants.*

IN THEIR OWN WORDS

My name is Sylva Joukalová, and I'm studying to be a biology teacher in Olomouc. The landscape of the Czech Republic has changed a lot since 1989; there are many more houses and roads being built. Olomouc itself is growing, and areas that were fields are now covered by supermarkets and stores. Places where there used to be garbage have been cleaned up and made into play areas for children. Now there are parks for children and special trash cans for people to clean up after their dogs. In the 1970s and 1980s the Krašhéhory Forest was almost destroyed by acid rain, from the prevailing winds that blew from the factories. After 1989 some of the factories were closed down and filters were put on the chimneys of those left open. Now the forest is flourishing again.

Mining

Although reserves are limited, black and brown coal is still produced in large quantities. The largest coalfields are in the northeast near Ostrava and in the far west near Sokolov. Open-pit mining methods are used to extract the brown coal, and this has a devastating effect on the landscape. Land that has been mined using the open-pit method is called "moonscape" because the land is pitted and bare like the surface of the moon.

▲ *Open-pit mining methods have scarred the landscape around Sokolov in the western part of the country.*

The Czech Republic has limited sources of the metallic ores iron, lead, and zinc. There is a source of gold just south of Prague, but mining there is banned because it would cause too much damage to the environment.

◀ *The western region of the Czech Republic provides a special type of clay, called kaolin. Here, it is being taken to a factory to be used to make porcelain and ceramics.*

Fish farming

Although the Czech Republic is landlocked, it produces significant amounts of freshwater fish. More than 23,000 lakes and ponds have been built for industrial fishing. The most important fish farming area is the Trebon district in the south of the country, where more than 70 percent of these ponds are located. The fish farmers use natural foods, rather than artificial feeding, to achieve better results in fish breeding. The common carp accounts for more than 85 percent of fish production. In Czech cuisine, breaded carp served with potato salad is an important part of the traditional Christmas meal.

▲ *Fillets of carp on display in a fish store. Most of the carp breeding ponds are in the Trebon district in the south of the country.*

◀ *Sport fishing is allowed in most areas if it is done at a certain time of year. All fishers must have a valid fishing license.*

Timber resources

The Czech forests and woodland cover one-third of the country. To make sure that forests are maintained, the number of trees being planted is greater than the number being cut down. Large areas of forest can be damaged in winter storms, when the wind is so strong that it breaks even big trees.

Most of the timber is softwood from trees such as pine or fir. Wood is an important material used in construction, furniture, and paper production. Experiments are being conducted using hemp fiber for paper production instead of wood, because hemp produces more paper than trees grown on the same area of land, and it takes less time to grow.

▼ *Trees that have been cut down are taken to the nearest railroad station before being loaded onto a train for further distribution.*

IN THEIR OWN WORDS

My name is Jiri Petrásek. At the moment I'm unemployed, but I used to be in charge of production in a timber yard that made doors and windows. Before the revolution the woods were owned by the state. Then, after 1989, they went back to private ownership, to people who didn't have any experience with managing forests. Some of them just wanted to make money, so they chopped down all of the trees and sold them, particularly for export. There was no control. Our company needed very big trees to work with, but now timber in the Czech Republic is too expensive even for the foreign market. So they buy it from Latvia and Lithuania. Now I'm hoping to get another job in a different timber yard, but it's hard for people in their fifties to find work.

Agriculture

Only 11 percent of Czechs are employed in agriculture. However, 40 percent of the land in the Czech Republic is used as farmland. The main plant products are grains (for making bread), potatoes, beets, hops, and fruit. Bread and potatoes are an important part of the traditional Czech diet.

◀ *Almost half of the Czech land is used for farming. The flat landscape is ideal for growing crops.*

▼ *This graph shows that the number of people who work in the agriculture industry has declined greatly since the 1960s.*

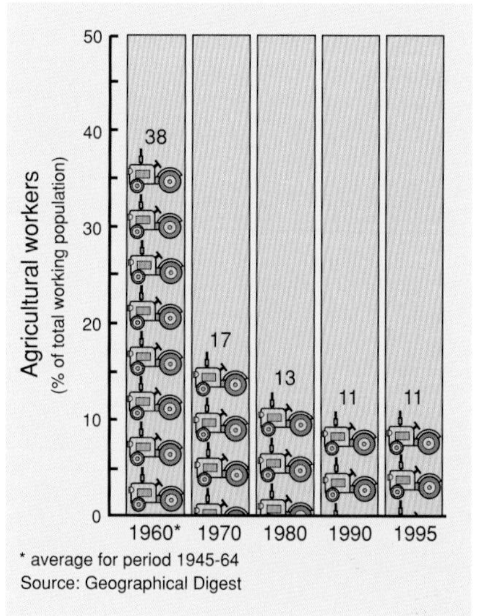

Hops are important for the brewing industry. At the beginning of the 20th century there were more than 600 breweries in the Czech Republic. However, many of the smaller companies have been taken over by competitors and there are now only about 80 breweries in the country.

South Moravia, in the southeast of the Czech Republic, is the warmest part of the country. Grapes that are good for making white wine are grown there.

In communist times, the Czech agricultural system used to run on the basis of collective agricultural teams. Each team was a self-sufficient farm. Every five years the government set production targets that determined how many crops were to be produced in a set period of time. The communist system was not very focused on protecting the environment, and, as a result, farmers used a lot of artificial fertilizers on the soil. Today, farming has changed from being government run to privately run. There is more focus on protecting the environment. Food grown on environmentally friendly farms is now becoming popular. People appreciate the difference in its taste and nutritional value, as well as its effect on the environment compared to food grown on a large scale using artificial fertilizers.

▲ *Many modern organic farms now keep fewer animals and focus more on their well-being.*

IN THEIR OWN WORDS

I'm Michal Polák. I'm currently doing my military service, but as soon as I finish I want to work on an organic farm. At the moment Europe is overproducing food; agriculture needs to be managed better. Some people don't support organic food since they prefer to pay for cheaper nonorganic products, so it's hard for our organic farmers to compete. There are two types of farms here: small-scale farms whose owners want to get the best from their land and sell in the local markets, and enormous intensive farms, the remnants of the communist regime, that damage the environment but supply supermarkets with cheap, plentiful food. I'm optimistic about the future because it's necessary to produce better food less destructively, and growing food organically does that.

5 The Changing Environment

Urbanization

Prague is one of the most quickly developing cities in the Czech Republic. Since the end of World War II (1945), the city has expanded to include many villages that formerly surrounded it.

Large-scale expansion of urban housing took place in many other Czech cities during the 1970s and 1980s. As a project of the communist government, thousands of people were housed in large apartment buildings called *panelaky*. Many of the *panelaky* were put close together in areas called *sidliste*, which means settlement. Most Czech people believe that these settlements are unattractive. Today, *panelaky* are not built anymore; people prefer to live in apartments or small family houses instead.

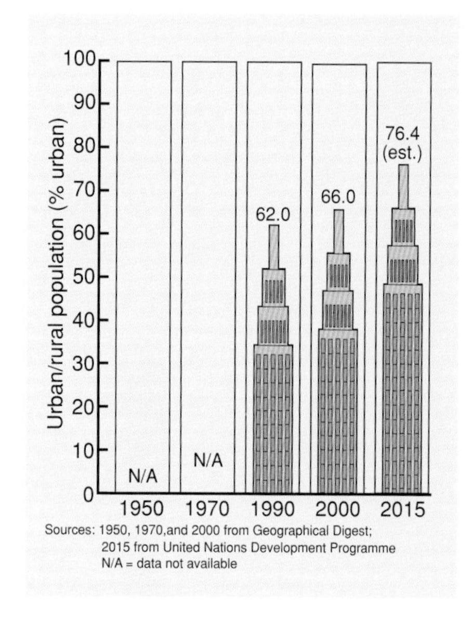

▲ *This graph shows that more than three-quarters of Czech people are expected to live in towns and cities by 2015.*

▼ *The infamous* panelaky *of the communist era are not being built anymore.*

Noise pollution

The increasing traffic on Czech roads produces a lot of noise, especially along highways and other main roads that heavy trucks use. In some areas tall barriers have been built to absorb the noise of the busy traffic.

Car ownership in the Czech Republic rose by 28 percent from 1988 to 1992. Then, from 1992 to 2003, this figure increased by another 30 percent nationwide. In Prague alone the number of cars owned has doubled in the same period.

In response to the increased number of cars, some cities have introduced electric-powered streetcars and buses to provide more public transportation. As a result, there are fewer cars on the roads and less traffic noise and air pollution.

▲ *Streetcars in Prague are a cheap and efficient way to get around the city.*

IN THEIR OWN WORDS

My name is Miehel Petrásek, and I work for a German freight company that transports goods all over Europe. Now that many people have access to the Internet, they are able to shop all over Europe, and companies are needed to deliver the goods. In the last few years our road system has grown. We are at the crossroads of Eastern and Western Europe, and our government knows it's important to build highways for the trucks that travel across the Czech Republic and other neighboring countries. But even with more highways, it is more difficult to get from place to place. Now there are a lot of traffic jams because more people own cars, and there are more trucks that slow down traffic.

Air pollution

The main sources of air pollution are the heat- and power-generating plants that burn fossil fuels, such as coal, and release massive amounts of harmful particles from their chimneys into the atmosphere. These gases form large brown clouds called smog. Smog is very harmful to the environment and to people. The exhaust from cars, trucks, and buses also contributes to smog pollution, especially in heavily populated areas with busy traffic. Air pollution on a global scale can cause a rise in Earth's temperature.

In the early 1990s, after the regime change, the government emphasized the need to reduce emissions. By the year 2000 the Czech program for the protection of the environment helped to reduce the levels of emissions by more than 50 percent compared with 1990 levels. The Czech Republic has also signed a number of international agreements on anti-pollution laws as a contribution to a cleaner environment worldwide.

▲ *A gas-powered plant producing heat for apartment buildings. Gas-powered plants cause less air pollution than those powered by coal.*

◀ *Money is being spent to restore buildings damaged by air pollution.*

Water pollution

Most cases of water pollution happen when sewage plants and factories release chemical and biological waste into rivers. Sometimes contaminated rainwater can soak through the ground to the rivers and spread pollution farther. This can make river and lake water dangerous to swim in or to drink. Certain fish species are also becoming endangered. The government is working hard to prevent dangerous waste leaks, and the levels of pollution have been decreasing since the late 1980s. However, the floods in 2002 were a setback. High water levels bring out oil from cars, flush the sewage systems into bodies of water, and can disperse dangerous materials from flooded chemical factories.

▶ *The government is tackling the problem of water pollution. Chemical pollution such as that shown here is on the decrease.*

IN THEIR OWN WORDS

My name is Roman Susanka. There are many forests here and everyone likes nature, but our environment was badly polluted by industrialization under communism. Now the government is trying to clean things up, but this is very expensive. Many people just care about themselves, but young people with a little more education care about the environment. In my school we have a biology club and we work on conservation projects.

Recycling

The amount of waste materials created in the Czech Republic is increasing. In 2001 the country produced nearly 45 million tons of garbage. It is becoming increasingly difficult to dispose of the garbage. A lot of trash gets dumped in landfills. Unfortunately, landfill leaks can pollute the soil and drain into underground water supplies. Another way to dispose of trash is to burn it in incinerators, but this causes air pollution. A good solution to the problem is to reuse waste materials in their recycled form. This requires separating waste products by type before they are collected for recycling. Specially designed containers for each type of waste have been introduced throughout the country. Disposing of waste in this way helps contribute to a better environment and reduces the cost of new products.

▲ *Illegal dumping still occurs in the Czech Republic despite changing attitudes toward environmental protection.*

◀ *Providing recycling bins in different colors makes it easier to sort trash for recycling.*

IN THEIR OWN WORDS

My name is Klára Volfová. Under the communist government the environment wasn't considered important. However, our country is now trying to improve the environment. Some of the ways it does this is by putting a limit on tree cutting and by planting more trees than are actually cut down. We can see many changes: before the revolution nothing was recycled, but now the government has given us different-colored bins for different materials so that we can recycle our waste materials and learn how to protect our environment.

Temelin power plant

The Temelin nuclear power station has been one of the major environmental safety concerns of recent years. The Russian- and U.S.-built reactor is supposed to be the best in its class. The government claims that the reactor is safe, although the opening has been repeatedly postponed because of failed safety checks, and experts say that it is not as safe as modern Western reactors. Following the tragic nuclear accident at Chernobyl in the former Soviet Union in the late 1980s, nobody wants to risk another disaster in the middle of Europe. The Temelin plant is one of two Czech nuclear stations. The other, Dukovany, is operating without problems.

▼ *Steam rises from the cooling towers of the Temelin nuclear power plant during a test program.*

6 The Changing Population

The population of the Czech Republic is currently slightly more than 10 million people. It had been growing slightly until 1993 when it reached its peak of just over 10.3 million. In 1993 the political changes of the late 1980s started to have a major effect on the Czech population. Before 1989, the average Czech couple had two children, which was more than in many Western European countries. By 1999 the number of babies being born had dropped by almost half. This sharp drop was also experienced by other European post-communist states, such as Hungary and Poland. The main reason for young Czechs having fewer children than in the communist past is a mixture of economic uncertainty and more freedom of choice. People feel less secure about the future than they did before 1989. They also choose to spend more time pursuing their careers rather than bringing up a large family. Because of the fall of the Soviet system, young people enjoy a wide choice of career, education, and travel opportunities. They are experiencing the kind of freedom their parents could only dream about. At the same time, they know that having children is a commitment that takes a lot of these options away.

▲ *Because the cost of living is rising, parents often decide that having just one child is sensible.*

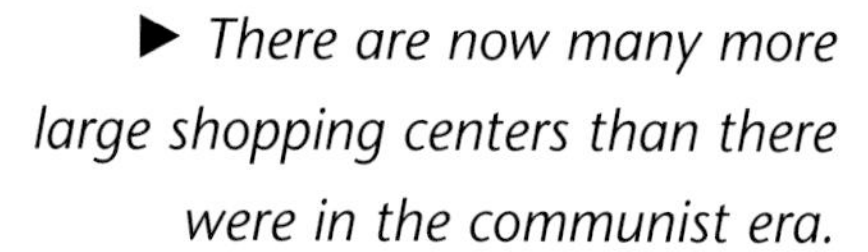

▶ *There are now many more large shopping centers than there were in the communist era.*

The majority of young Czech people expect to live a lifestyle similar to that of people in the United States or England, but they can't always afford to do this. Many couples decide to have only one child. If this trend continues, the Czech Republic will have a higher number of elderly people. The government expects the situation to stabilize by 2010, but this depends on the country's economic and political position in Europe.

▶ *The number of elderly people in the Czech Republic is on the increase. This grandfather is looking after his grandchild while the parents are at work.*

IN THEIR OWN WORDS

My name is Milan Fait. I travel around the country selling pharmaceutical (medical drug) products. A big change here is that more people are getting divorced. Before the revolution it was considered bad for couples to split up, but now women have well-paid jobs so they can divorce and bring up children by themselves if they choose. My parents have been married for more than 30 years and I'm very proud of that, but I know a lot of young people who have divorced after only a few years. People are also having fewer children now and having them when they are older. My parents were in their twenties when they had me, but now women have children in their mid-thirties.

Ethnic groups

The Czechs currently represent 94 percent of the country's population. When the Czech Republic split from Slovakia in 1992, some Slovaks remained and now represent roughly 3 percent of the population. The other 3 percent are mainly Polish, German, Roma, and other ethnic groups. Most of the Roma people (also known as gypsies) are Czech natives, but they still use their own unique language.

▲ *The Vietnamese have been very successful in the marketplace, selling mainly tobacco, alcohol, and clothes to tourists.*

Immigration

Before 1989 it wasn't as easy to travel around Europe as it is today. The political differences kept Eastern and Western Europe well separated. Currently, there are large numbers of immigrants who travel across Europe from east to west to find a better life for themselves and their families.

People from poorer countries in Asia and Eastern Europe, such as Vietnam, Turkey, Ukraine, Moldova, Romania, Russia, and Bulgaria, travel to escape war or to seek political asylum in more stable and peaceful countries. Before 1989, people came to former Czechoslovakia from communist Vietnam as a part of a cooperation plan among communist countries. Many of the immigrants decided to stay and adopt Czech citizenship.

▶ *A Roma woman in Olomouc sells magazines. The Roma community now has its own advisers who consult with the national and local governments.*

In Prague more than 5 percent of the inhabitants are foreign citizens. Many are employees of the capital's foreign embassies and headquarters of companies that have business in the Czech Republic. Some immigrants have no money or travel documents and try to cross the border illegally. The Czech Republic is becoming an increasingly popular place for immigrants from the East, mainly because its central position in Europe makes travel to other countries more convenient.

▶ *A Nigerian student sells cruise tickets in Prague. The high standard of education in the Czech Republic attracts many foreign students, who take advantage of the relatively cheap university fees.*

IN THEIR OWN WORDS

My name is Janis Tsolakidis. I am a student from Cheb. My mother's parents were from Slovakia and my father's from Greece. My grandfather was a communist in Greece and when they lost the civil war there he had to flee here. There are quite a few Greek families who came at that time. Several Vietnamese children go to our school. It's difficult when there are several in a class because they speak Vietnamese with one another and don't join in with us. It doesn't matter what people look like; what's important is how they think and behave.

7 Changes at Home

Family life

More than 70 percent of Czech people live in apartments in urban areas. The typical family has only one or two children. Because there is a housing shortage, most young adults continue to live with their parents until they can afford their own place. It is also common for the grandparents to live with their family. The grandmother, *babicka*, often plays an important family role and is looked on with respect and love. A grandmother often lives with her children and looks after her grandchildren when both parents are at work.

▲ *Most Czech people live in apartments, which are part of buildings like these. Many that were built decades ago have been repainted and modernized to appeal to potential buyers.*

IN THEIR OWN WORDS

I'm Jovana Janíková, and I am a construction engineer and building inspector in Olomouc. Although there are many women engineers, they usually earn less money than men do. The main role of women in the Czech Republic is still to take care of the family. Many employers are afraid to hire women who have children in case they need to take time off work to care for their family. I think life is easier for single women who can build their careers, especially in large cities. Child care depends on the financial situation of the family; it's common to arrange for grandparents to look after children. I work freelance so I can take time off for my children.

◀ *Military service is compulsory for all eighteen-year-old men. Most men complete this service before they marry.*

▼ *This graph shows that in recent years the population in the Czech Republic has begun to decline.*

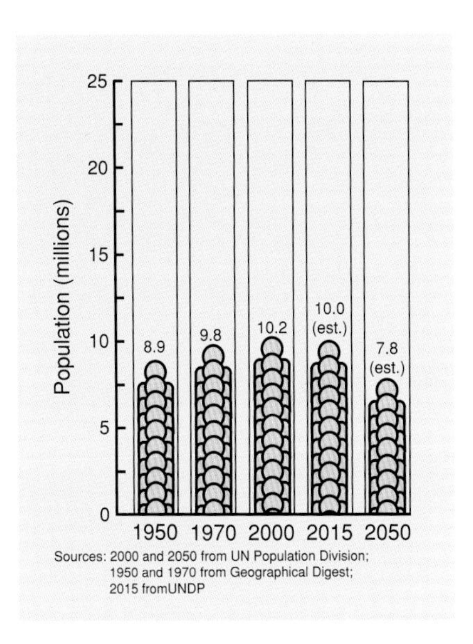

In the past, Czechs tended to marry young, especially in rural areas. Young men often married before they began military service at the age of eighteen. In communist times military service lasted two years, but now it is just one. For young men who don't want to train in the army for personal reasons, they can do alternative civilian service for eighteen months, which involves various non-military jobs in public service. Today, people are marrying later in life. Some people have traditional village weddings with a lot of singing, dancing, and elaborate meals.

Many Czechs who live in cities also have weekend homes in the country, where they have gardens and enjoy the outdoors with their families.

Religion

The most widespread religion in the Czech Republic is Roman Catholicism; currently 40 percent of the population is Catholic. Catholicism is very popular in the Moravia region in the east of the country, where people of all ages attend church regularly. Pope John Paul II visited Moravia in 1990.

▲ *Catholicism is the most commonly practiced religion. This picture shows St. Vitus Cathedral in Prague, the largest church in the country.*

The Czech Republic also has a well-established Jewish community that has several thousand members. A synagogue in Prague has the names of more than 80,000 Czechoslovakian Jews who were killed in the Holocaust of World War II (1939–1945).

A large percentage of Czech people say they are atheist (about 40 percent). This is partly because the communist regime discouraged religion. The Communist Party leaders believed that religious people would be less likely to follow the socialist doctrine.

▶ *One of the two active synagogues in Prague. Many tourists from other countries come to see the synagogues.*

Education

The level of literacy in the Czech Republic is high, at 99.9 percent. Primary and secondary education in the Czech Republic is free for all children. Children start at the age of six and finish at fifteen, when they can choose to go on to higher education. Each primary school class has its own class teacher who is responsible for attendance and conduct. After the first two years, students have different teachers for different subjects. Higher education used to be free in the communist era, but new economic conditions mean that students now have to pay for most of their college or university education. Another problem students face is finding housing. In the past that, too, was organized by the communist government.

▼ *Students in Cheb leave their secondary school in the afternoon. Most classes start at 8:00 a.m.*

IN THEIR OWN WORDS

My name is Jitka Jurácová. I work in the university library at Olomouc. I like reading and wanted to work with books. I took a special one-year course for librarians, and then I worked in a bookstore for about ten years. Before the 1989 revolution, people were not allowed to read many books. Afterward, books by Czech writers who had been prohibited were published. People wanted to read more and more and to learn languages like English, French, and German. For three or four years, there was a big increase in reading and publishing. Before 1989 there were only three or four bookshops in Olomouc. After the revolution there were about thirty. Now, however, there are about six, because many people would rather watch television than read. Many children are more interested in computers and computer games than in reading.

Changes in travel opportunities

In the communist era the government subsidized the leisure activities of workers. People could easily afford to spend a nice vacation with their family, but only in countries that were also members of the Eastern Bloc, such as Bulgaria or Hungary. Now there are no such restrictions. People can pursue a variety of interests.

The increase of travel options has brought about a boom in the service industry. For example, before 1989 there was just one principal travel agency for the whole country. Older Czech people are now able to visit countries they could only dream of before. Traveling has also quickly become very popular with the younger generation, and today you can meet young Czechs who are traveling all over the world.

▲ *The cost of traveling abroad has decreased since 1989. The Czech national airline now flies people all over the world.*

◀ *Young people in the Olomouc railroad station. The price of tickets has gone up steeply in the past few years.*

Sports

Czech people are very fond of sports all year round. The most traditional sports are soccer and ice hockey. The Czech Republic ice hockey team is one of the best in the world. With the increase of Western influence after 1989, many new sports were introduced, and the Czechs are always ready to try them. Skateboarding became very popular with the younger generation, and snowboarding has offered an exciting alternative to skiing. Other popular sports include tennis, volleyball, and ice skating.

► *Ice hockey is one of the most popular sports in the Czech Republic. Many people support the national team.*

IN THEIR OWN WORDS

My name is Linda Novakova, and I live in Cheb. In my free time I enjoy playing sports, going to a café for a coffee, or watching television or a video. When my mother was young she spent her free time horseback riding. She lived in a small village. Life was very different for her as a girl because she had to help more in the home and garden. Sometimes I work as a camping instructor with children from the city. I love doing that. I also love traveling; I've been to Germany, Spain, and Greece. We have a small chalet in the mountains and I go there on vacation.

Diet

The diet and eating habits of the Czechs have begun to change because there is now a wider choice of foods and people are becoming aware of health issues. However, traditional foods are still very popular.

The most popular option for eating out is still in hotels, bars, and restaurants. Traditionally, a large number of meat dishes appear on menus. Eating meat, especially pork, is very popular in the Czech Republic. Many Czechs enjoy roast pork with pickled cabbage and sliced-up dumplings, or stewed beef goulash.

▲ *Restaurants are a big business in Prague, where there are always enough people who like to eat out.*

Since the changes in 1989, however, new influences and trends have appeared. New recipes and exotic foods are now available. There are several specialized tea shops in Prague, where you can buy different teas from around the world. Although the diet of the average Czech hasn't dramatically changed, there are some definite signs of an increasing focus on healthful eating. Health-food stores that sell organically grown produce are becoming popular.

◀ *Fast food, like these hot dogs, is increasingly available and popular, although people are now becoming more aware of fast food's negative impact on their health.*

IN THEIR OWN WORDS

I'm Hana Rudorferová. I have a store that sells organic foods, herbs, and medicines. My store has been open for ten years, and I'm doing very well. Since the revolution in 1989, people have become more interested in health foods. Before this no one knew about organics and health food stores. Eating habits, diet, and people's attitudes toward life and free time have all changed enormously. We live much healthier lives now, even though we work harder. Many people like me exercise every morning, eat less meat, and drink a lot of water. In the past, very few people were vegetarian, but now it's widespread, especially among women.

Health care

In the past, the quality of health care in the Czech Republic was very good compared with that of other European countries. In the communist era the state paid for all health care. Today people have to pay for health insurance, except for children, the retired, and the unemployed. After 1989 many hospitals and clinics were privatized and became responsible for their own financing. Private clinics charge patients when they come to see a doctor.

Currently, the Ministry of Health is struggling to keep the quality of health care affordable, because medical costs are getting more expensive and the government is reluctant to increase spending on medical care.

◀ *A private clinic in Cheb. Private hospitals started functioning after 1989 and have been providing a good level of health care under the control of the Ministry of Health.*

8 Changes at Work

Unemployment

Under communist rule people were guaranteed a job. Because of this it was very easy to get work, even for people with no skills, references, or qualifications. This helped to keep unemployment very low. The only social benefits that people took advantage of were maternity leave and sick pay, which were both paid for by the state. Today jobs are much less secure and levels of unemployment are rising.

▲ *Job security has decreased since 1989 and people have to make more of an effort to find work.*

One of the reasons it is more difficult to get a job today is the situation in the housing market. The prices of houses are high, and it is not easy to find a suitable home that is close to work, so people struggle to find employment close to home. Some of the regions most affected by rising unemployment are in the north of the country, where a lot of people work in the coal-mining industry. When the miners worked under the communist regime, they were paid high wages. Now the miners still receive a good salary, but

▼ *Because many people are now coming to bigger towns looking for work, housing is becoming more of a problem, especially in cities such as Prague.*

IN THEIR OWN WORDS

My name is Michal Kovar. Before the revolution it was very hard to do what you really liked. You had to know the right people in the right places, and work was much less flexible. Now it's easier to try something and, if it doesn't work out, to change. Now it's up to you what you do. I studied psychology but then gave that up and worked in a bookstore. Then I became interested in photography and am now writing reviews of digital cameras. The problem before the revolution was that people didn't work so hard because they didn't benefit if they did. People expected the government to look after them. Many people are not used to running their own lives; they expect someone else to come and solve their problems. It will take time for people to get used to the new freedom.

compared with what they used to earn their wages have been reduced and they are reluctant to take less well-paid jobs. Experts say that what is needed to help the regions hit by unemployment is making them more accessible to investors from abroad. This would allow more businesses to develop, and the number of job opportunities would increase.

◀ *These streetcar workers are fixing a rail junction. Jobs for the city are usually secure and well paid.*

Tourist industry

Many people now work in the tourism industry. After 1989, Prague quickly became the most visited capital of central Europe. Since then, Prague's reputation has grown and the Czech Republic has become well known for its culture and history.

The increased demand for tourist hotels has also boosted the Czech hospitality industry. The quality and choice of hotels has improved since 1989.

It isn't just Prague that people come to see, though. Many visitors are attracted to the beautiful scenery throughout the country, especially its wonderful architecture and well-kept castles. The western spa region is very popular with tourists for its abundance of cold and hot mineral springs, many of which are said to have healing properties. The Czech Republic also offers affordable opportunities for visitors from Western Europe to pursue winter sports. Major ski resorts are located in the Krkonose Mountains in the north.

▲ *The Presidential Palace in Prague attracts many tourists from around the world.*

◄ *There are many stores in Prague that sell traditional goods to tourists, such as this puppet shop.*

IN THEIR OWN WORDS

I'm Karel Triska, and I'm eighteen. The biggest change for us is that now we can travel abroad. Before we could only go to countries like Hungary, the Soviet Union, and Bulgaria, but now we can go everywhere.

In my school we use a lot of computers. A few years ago the government passed a law that all schools, primary and secondary, should have free Internet and computers. My school has two rooms with about 45 computers for 1,000 students ages 12 to 20. When I finish school, I hope to work with computers.

Computer industry

Until the early 1990s, computers were available to only a few businesses. Now the number of people who have a computer at home and work has increased dramatically with the arrival of the Internet. Many new computer-related businesses have emerged, including computer gaming and Internet cafés. People have quickly adapted to computers, and they are becoming a part of everyday life.

▶ *This café offers public Internet access. The Internet has become more available since the late 1990s.*

Changes in the economy

The Czech economy has recently undergone major transformations. Soon after the end of World War II it became obvious that Czechoslovakia was required to follow economic steps approved by the Soviet Union. After 1955 the Czech Republic could no longer do business with the West.

Before 1989, almost all companies were owned by the state, which is typical for communist countries. A transformation to a Western-style economy was needed before the Czech Republic could function economically within Europe. A new stock market was established so that the shares of the privatized companies could be traded. Every person over the age of eighteen received a share of vouchers from the government. With these vouchers people were able to buy shares in the privatized companies.

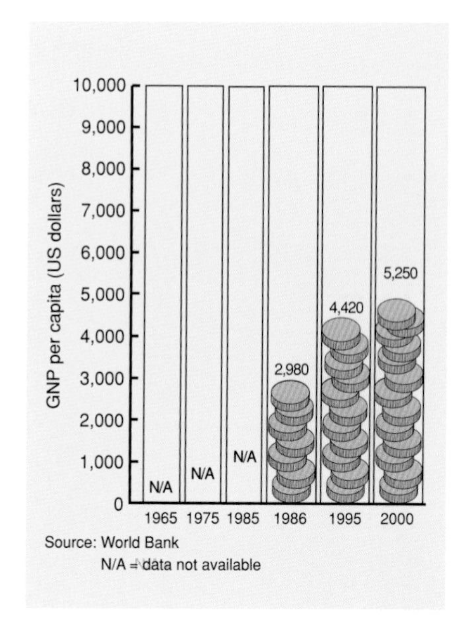

▲ *This graph shows that since 1985 the GNP (gross national product) has risen steadily.*

▼ *Since 1989 the ways that people handle their money have changed a great deal. For example, instead of going to the bank to withdraw money people can now take it out at ATMs.*

IN THEIR OWN WORDS

I'm Martin Hladecek and I have my own business. I travel around the country selling gas appliances and water heaters. Now we have more opportunities to travel and build our own careers. Businesses have contacts with the rest of the world. Life is much better now. I think of myself as a European and travel to other countries a lot; cheap flights have made this much easier. Our republic is now visible and we are firmly on the map of Europe; in fact, we are at the very heart of Europe.

The Skodovka

The car manufacturer Skoda is a major Czech exporter. The factory, which is known as "Skodovka," is situated in Mladá Boleslav in the north of the country, and today it still employs thousands of people. The company is a good example of the transformation of Czech industry that began after 1989. The recent political and economic changes in the Czech Republic have had a great effect on the company.

As a part of the Soviet bloc, the Skoda factory produced low quality cars that sold to the masses for a cheap price. In 1991 the company was bought by their German counterpart, Volkswagen, and today it produces quality cars that successfully sell in many other countries, including Germany, France, and the United Kingdom.

▼ *Because of dramatic improvements in quality, Skoda remains the best-selling car in the Czech Republic, despite competition from abroad.*

Women at work

Women and men in the Czech Republic, and previously in Czechoslovakia, always had equal opportunities in education and employment. Today, Czech women pursue higher positions much more often than they did in the past. More women today have jobs as executives in companies or for the government, which was not common in the communist state. Although many women work full-time outside the home, they are also responsible for housework and raising children. The state continues to provide generous maternity leave and social benefits for mothers, especially if they are single or have financial difficulties. The ever-increasing pace of life and career options have led many Czech women to decide to have children later in life so that they can concentrate on their career first.

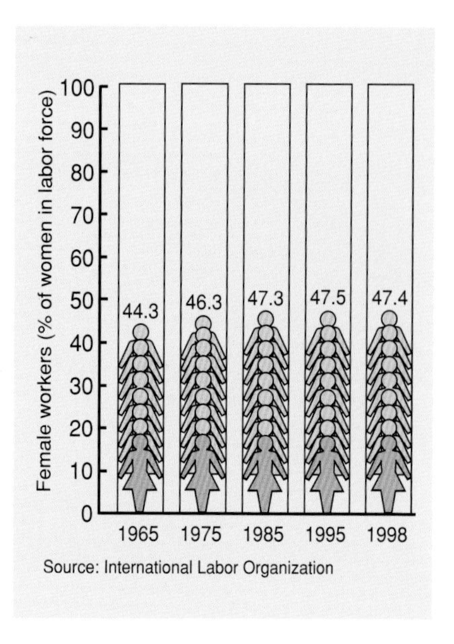

▲ *This graph shows that although the number of women in employment has remained constant since the 1960s, there are more women in top jobs now than before.*

◀ *A woman surveyor at work. Czech women have equally good job opportunities as men.*

Freedom in art

The Czech art scene prospered greatly from the change of regime in 1989. Artists today enjoy real freedom, whereas in the communist era the government always had control over what an artist produced. Any art that was not in accordance with the communist idea could result in artists being prosecuted for betraying their country. Many artists who wanted to express themselves freely were imprisoned for undermining the government's authority. Among the many who refused to restrict their creativity was the playwright Václav Havel. He was imprisoned for many years. After the revolution Havel became president of the Czech Republic.

▲ *Today's government does not put any restrictions on freedom of speech, so musicians can sing about what they like.*

IN THEIR OWN WORDS

I'm Lukás Kosek, and I work as a freelance visual designer for television commercials. Over the last ten years, television advertising has changed a lot. Now we look to the West for our influences. Television advertising is now more professional and many foreign companies come here because it is cheaper.

After school I began to work in an advertising agency doing graphic design. I didn't continue with further education because I knew what I wanted to do. Two years ago I came to Prague and worked at the film academy. I met a lot of people there who helped me get work.

9 The Way Ahead

Between the two world wars (1918-1939), Czechoslovakia was a well developed country in Europe, both culturally and economically. The events of World War II resulted in the political division of Europe. Although the western part of Czechoslovakia was liberated from German occupation by the U.S. army in 1945, soon afterward the whole country was firmly in the grip of the Soviet-controlled communist government. The country was occupied by the Russian armies from 1968 until the fall of the communist regime in the late 1980s. Under this repressive government many Czechs gave up trying to change things because it seemed like the Soviets would be in power forever.

▲ *Some traditions continue, for example, the changing of the Guard at Prague Castle.*

When the regime changed in 1989, people were surprised. The rapid changes in the economy that followed were a result of the effort to build up trade with the West again. This really put the new government to the test. The massive amount of state-owned capital was transformed into private businesses. This gave rise to many new businesses, and foreign investors took the opportunity to join in.

Many older people have forgotten about the past disadvantages of the communist system and are disappointed

▶ *The Dancing House is an example of Prague's modern architecture.*

◀ *Most young students are confident about their chances of success in the future.*

because everything has become so expensive. However, before 1989, when people wanted to buy something they had to wait in a long line and even then might not get what they wanted. Overall, the Czech people find the new system much better. The freedom to do or say what they like cannot be measured by money. Despite various setbacks, the prospects for the economy are good and there are more opportunities for an exciting and varied life.

IN THEIR OWN WORDS

My name is Tereza Hercigová. I hope we will join the European Union and that things will improve. It will give us more opportunities to live and work abroad. Some people are afraid that prices will go up and are suspicious of foreigners. Communism was very bad for this country because everything belonged to the state, so no one really cared about the stores and factories. It didn't matter how well or badly anyone worked because they were paid anyway. This attitude needs to change and people have to work harder. If we do, our future will be bright.

Glossary

acid rain heavily polluted rain of high acidity that is very harmful to the environment

Adriatic Sea part of the Mediterranean Sea between Italy and the Balkans

Baltic Sea body of water lying between the coasts of Germany, Poland, and Scandinavia

communism political system in which the government controls all production and in which there is no private property

Eastern Bloc countries of Eastern Europe that had communist governments and supported the Soviet Union

ecological describing the relationship between people and the environment

emissions gases or other substances that are released into the air

environmentalist person who is engaged in activities aimed at the protection, preservation, and research of the natural environment

federation system of government in which two or more states unite into one country

hydropower electricity generated by turbines that are turned by the force of falling water

Holocaust mass killing of Jews, Roma (gypsies), homosexuals, communists, and others in Europe by the Nazis in the 1940s

Moravia eastern region of the Czech Republic where people speak a distinct dialect of Czech

North Sea body of water that lies to the north of Europe

organic food food grown without the use of artificial methods and fertilizers and therefore contains few, if any, harmful ingredients

soviet elected government council in a Communist country

subsidy money that is paid by a government to reduce the cost of producing goods

synagogue building where Jewish people meet for religious worship, also known as a temple

Further Information

Books

Humphreys, Rob. *Country Insights: Czech Republic.* Chicago: Raintree, 1998.

Nigel, Ritchie. *Communism.* Chicago: Raintree, 2001.

Nollen, Tim. *Czech Republic.* Milwaukee: Gareth Stevens, 1999.

Sioras, Efsthathia. *Czech Republic.* Tarrytown, N.Y.: Marshall Cavendish, 1999.

Useful Addresses

Embassy of the Czech Republic
3900 Spring of Freedom St., N.W.
Washington, D.C. 20008
(202) 274-9100
www.mzv.cz/washington

Embassy of the United States in the Czech Republic
Triziste 15
118 01 Prague 1
www.usembassy.cz

Agency for Nature Conservation and Landscape Protection of the Czech Republic
Kalinsnická 4-6
130 23 Prague 3
email: apok@nature.cz

Index

Page numbers in **bold** refer to pages with photographs, maps, or statistics panels.